水火“相容”

——镜泊湖的地质故事

“COMPATIBLE” WITH FIRE AND WATER

—GEOLOGICAL STORY OF JINGPOHU

李江风 孙德志 马晓群 周学武 付崇华 万沙 陈梦婷 田野 编
Edited by Li Jiangfeng Sun Dezhi Ma Xiaoqun Zhou Xuewu Fu Chonghua Wan Sha Chen Mengting Tian Ye
胡志红 译
Translated by Hu Zhihong

中国地质大学出版社
ZHONGGUO DIZHI DAXUE CHUBANSHE

空山新雨后 Mountains After the Rain
摄影：吕国辉 Photographed by Lü Guohui

目录 CONTENTS

01 公园简介

BRIEF INTRODUCTION

镜泊湖世界地质公园位于中华人民共和国黑龙江省东南部宁安市境内，牡丹江流域中上游。公园地理坐标为东经128°30′00″—129°11′00″，北纬43°43′34″—44°17′55″，总面积1400km²。

Jingpohu UNESCO Global Geopark is located at Ning'an City in the southeast of Heilongjiang Province. Situated at 43°43′34″—44°17′55″ N and 128°30′00″—129°11′00″ E, in the middle and upper Mudanjiang reaches, it covers a total area of 1400 km².

镜泊峡谷 Jingpo Valley
镜泊湖世界地质公园管委会提供 Provided by Jingpohu UGGp Management Committee

2005年8月，镜泊湖地质公园被国土资源部(现为自然资源部)批准为国家地质公园。2006年9月，经联合国教科文组织批准，镜泊湖国家地质公园正式成为世界地质公园网络成员。

镜泊湖世界地质公园是以火山口及其地质地貌和火山熔岩堰塞湖为特色，融合森林湿地等自然生态景观、历史文化遗址和现代文明等人文景观于一体的特大型世界地质公园。

镜泊湖世界地质公园是研究中国第四纪火山活动的最佳基地之一。这里有世界上最大的熔岩气洞塌陷型瀑布——吊水楼瀑布，世界上最大的火山熔岩堰塞湖——镜泊湖，国内外罕见的典型火山熔岩喷气锥和喷气碟，规模宏大的地下多级熔岩隧道以及保存完好的火山锥群等各类火山地质遗迹景观。

绚丽多姿的自然风光、健康完整的生态系统使镜泊湖保持了人与自然的高度和谐。这里有生长在火山岩“石板”上的响水大米，有东北虎、丹顶鹤、中华秋沙鸭、红松、东北红豆杉等国家级保护动植物，有气势恢宏的渤海国遗址，还有独具特色的朝鲜族、满族等少数民族村落。历史与现实在此汇聚，自然与人文在此交融，使得镜泊湖世界地质公园焕发出勃勃生机。

镜泊湖如同一颗璀璨夺目的明珠，镶嵌在中国的北疆之上。这里春天山花烂漫，夏天翠绿成荫，秋天红叶似火，冬天白雪皑皑。百里长湖，山峦叠嶂，鸟语花香，风光无限。

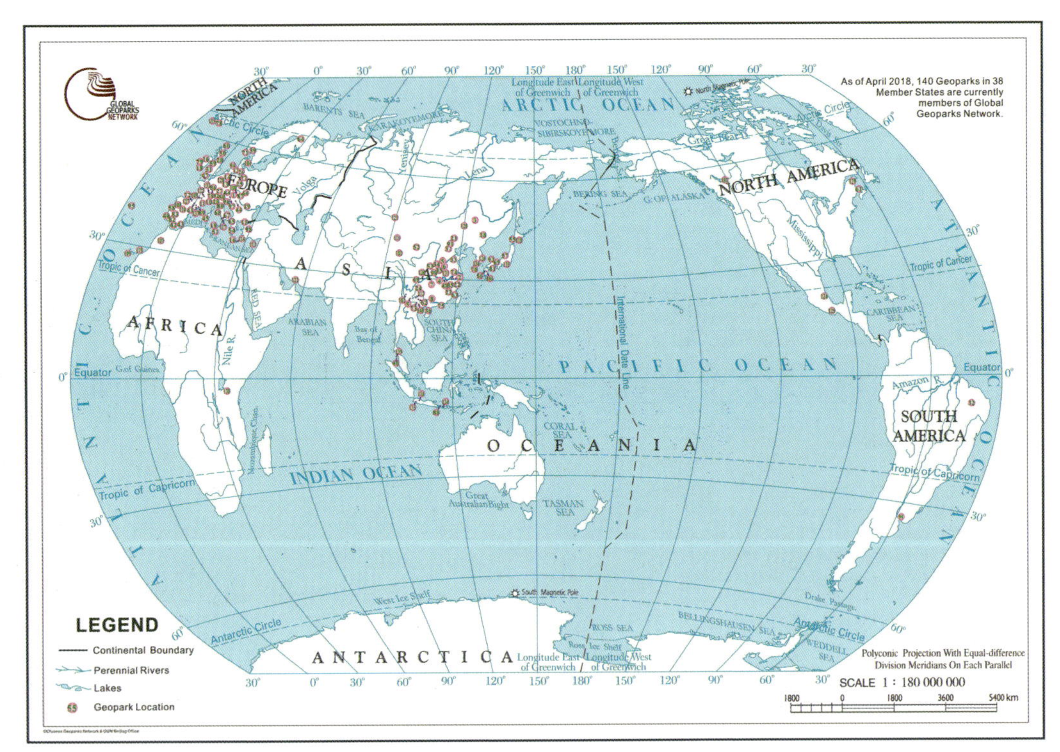

全球世界地质公园分布 Distribution of GGN Members

镜泊湖地质公园在黑龙江省的位置
The location of Jingpohu UNESCO Global Geopark in Heilongjiang Province

In August 2005, Jingpohu Geopark was approved as National Geopark by Ministry of Land and Resources (it has been changed into Ministry of Natural Resources now). In September 2006, it officially joined in the Global Geoparks Network after getting the permission from UNESCO and has renamed Jingpohu UNESCO Global Geopark (UGGp) now.

Jingpohu UGGp is an extra large-sized Global Geopark features volcanic craters, geological landforms and lava dammed lake, integrating natural ecological landscape such as forest, wetland, and humanistic landscape such as historical and cultural sites and modern civilization sites.

Jingpohu UGGp is one of the best bases for studying the Quaternary Volcanic Activity in China, here scatters various volcanic geological relics landscapes, such as Diaoshuilou Waterfalls, known as the world's largest lava collapse crater waterfalls, Jingpo Lake, known as the world's largest volcanic lava dammed lake, typical volcanic lava fumarolic cone and fumarolic dish rare both in China and at abroad, massive underground multistage lava tunnels and well-preserved volcanic cones.

The colorful and beautiful natural scenery as well as the healthy and entire ecosystem has kept a high degree of harmony between man and nature in Jingpohu UGGp. The world-famous Xiangshui Rice grown on the volcanic basaltic is produced in this land. There are national protected plants and animals such as *Panthera tigris altaica*, Grus *japonensis, Mergus squamatus, Pinus koraiensis* and *Taxus cuspidate*. There are also magnificent Relics of Ancient Bohai kingdom, and distinctive villages of Korean, Manchu and other ethnic minorities. Thanks to the perfect integration of history and reality as well as the impeccable fusion of nature and humanity, Jingpohu UGGp is full of vigor and vitality.

Jingpohu is a dazzling pearl in the northern China, with every season having a distinct beauty and creating a delightful view. In spring, it features a myriad of blossoms unique in the mountains. In summer, it boasts serene forests and dense foliage. In autumn, flaming red leaves ignite its landscape. In winter, it even more captivating in glittering white snow. The Long Lake meanders through the lush rolling mountains, showcasing the idyllic paradise on earth with enchanting scenery highlighting twittering, birds and fragrant flowers.

镜泊湖鸟瞰 Aerial View of Jingpo Lake 镜泊湖世界地质公园管委会提供 Provided by Jingpohu UGGp Management Committee

景点 · SCENIC SPOTS

01. 火山口森林复火山 ········· Forest in Crater Composite Volcanic Landform
02. 火山碎屑物 ········· Pyroclastis
03. 火山渣 ········· Cinder/Scoria
04. 浮岩 ········· Pumice or Scoria
05. 火山弹 ········· Volcanic Bomb
06. 炭化木 ········· Carbonized Wood
07. 雄狮岩洞远观 ········· Overlook of Lion Cave
08. Ⅳ号火山口 ········· Crater No. Ⅳ
09. 坐井观天 ········· Looking Up to The Crater
10. 多期次火山熔岩 ··· Lava Flows of Multiple Eruptions
11. 雄狮岩洞 ········· Lion Cave
12. 千层岩 ········· Multi-layered Rocks
13. Ⅲ号火山口 ········· Crater No. Ⅲ
14. Ⅱ号火山口 ········· Crater No. Ⅱ
15. Ⅰ号火山口 ········· Crater No. Ⅰ
16. 生命之路 ········· Road of Life
17. 齐天亭 ········· Qitian Pavilion
18. 鸟语林 ········· World of Birds
19. 松拜佛 ········· Pine Bowing to Buddha
20. 鸳鸯池 ········· Mandarin Duck Pool
21. 响泉 ········· Echo Spring
22. 地下熔岩瀑布 ········· Underground Lava Cascade
23. 古冰洞 ········· Ancient Ice Cave
24. 熔岩洞 ········· Lava Cave
25. 威虎厅熔洞 ········· Tiger Hall Lava Cave
26. 洞中洞 ········· Cave in Cave
27. 神羊洞 ········· Shenyang Cave
28. 小北湖 ········· Xiaobei Lake
29. 钻心湖 ········· Zuanxin Lake
30. 镜泊石海 ········· Jingpo Lava Stone Sea
31. 紫菱湖 ········· Ziling Lake
32. 塌陷的熔岩丘 ········· Collapsed Lava Dome
33. 喷气碟 ········· Fumarolic Cone
34. 吊水楼瀑布 ········· Diaoshuilou Waterfalls
35. 博物馆 ········· Museum
36. 抗联纪念碑···Anti-Japanese War Memorial Monument
37. 邓小平题词碑 Inscription by Premier Deng Xiaoping
38. 渤海风情园 ········· Balhae Garden
39. 红罗女文化园 ········· Hongluonv Cultural Garden
40. 靺鞨绣坊 ········· Mohe Embroidery Workshop

景点 · SCENIC SPOTS

41. 城子后山城 ········· Chengzihoushan City
42. 镜泊峡谷 ········· Jingpo Valley
43. 奇径碑园 ········· Qijing Stele Garden
44. 叶剑英题诗碑 ··· Inscription by Marshal Ye Jianying
45. 毛公山瞻仰台 ········· Platform for Viewing Chairman Mao Mountain
46. 药师古刹 ········· Pharmacist Temple
47. 少奇木屋 ········· Liu Shaoqi Cottage
48. 少奇钓鱼台 ········· Liu Shaoqi Fishing Platform
49. 镜泊小镇 ········· Jingpo Town
50. 漱玉潭 ········· Jade Pool
51. 镜泊湖 ········· Jingpo Lake
52. 重唇河山城 ········· Chongchunheshan City
53. 鹿苑岛 ········· Deer Island
54. 毛公山 ········· Chairman Mao Mountain
55. 白石砬子 ········· White Rock Cliff
56. 大孤山 ········· Dagu Mountain
57. 小孤山 ········· Xiaogu Mountain
58. 镜泊湖边墙 ········· Jingpohu Wall
59. 东方净琉璃世界 ········· Eastern Glass World
60. 湖州城 ········· Huzhou Town
61. 珍珠门 ········· Pearl Gate
62. 道士山(三清观)··· Taoist Mountain(Sanqing Temple)
63. 阎王鼻子 ········· Yama's Nose
64. 老鸹砬子 ········· Crow Mountain
65. 抗日纪念碑、翰章园 ··· Monument of Anti-Japanese War and Chen Hanzhang's Tomb
66. 抗日墙缝战场遗址 ··· Ruin of Qiangfeng Battlefield
67. 莺歌岭 ········· Yingge Hill
68. 熔岩台地 ········· Lava Plateau
69. 兴隆寺 ········· Xinglong Temple
70. 上京龙泉府遗址 Relics of Captial Shangjing Longquanfu

常用电话 · USEFUL TEL

咨询电话(Info Tel)········· 0453-6270180
投拆电话(Complaints Tel)········· 0453-6270080
急救中心(First-aid Centre)······ 0453-6270019/120
报警电话(Emergency Tel)······ 0453-6270000/110
火山口森林(Crater Forest)········· 0453-5777779
观光车公司(Tourism Bus)········· 0453-6270111
观光船公司(Tourism Boat)········· 0453-6270055

火山口森林导览图
GUIDE MAP OF FOREST IN CRATER

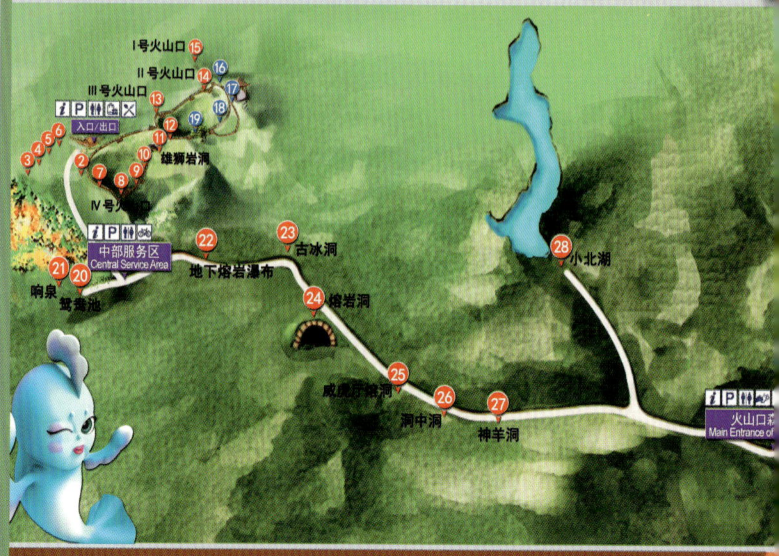

镜泊湖导览图
GUIDE MAP OF JINGPOHU LAKE

路线信息
SIGHTSEEING ROUTE

步行路线 Walking Route

镜泊湖北门(North Gate of Jingpo Lake) (750m/10min) → 吊水楼瀑布(Diaoshuilou Waterfalls)
吊水楼瀑布(Diaoshuilou Waterfalls) (3.7km/40min) → 镜泊山庄码头(Jingpo Villa Wharf)
镜泊湖东门(East Gate of Jingpo Lake) (4.5km/50min) → 吊水楼瀑布(Diaoshuilou Waterfalls)
镜泊湖东门(East Gate of Jingpo Lake) (5.3km/1h) → 镜泊湖北门(North Gate of Jingpo Lake)

扫描进入

观光车路线 Tourism Bus Route

镜泊湖北门(North Gate of Jingpo Lake) (750m/2min) → 吊水楼瀑布(Diaoshuilou Waterfalls)
吊水楼瀑布(Diaoshuilou Waterfalls) (3.7km/8min) → 镜泊山庄码头(Jingpo Villa Wharf)
镜泊湖东门(East Gate of Jingpo Lake) (4.5km/10min) → 吊水楼瀑布(Diaoshuilou Waterfalls)
镜泊湖东门(East Gate of Jingpo Lake) (5.3km/12min) → 镜泊湖北门(North Gate of Jingpo Lake)

扫描获取电子

观光船路线 Tourism Boat Route

全湖(Long Route)：镜泊山庄码头(Jingpo Villa Wharf) 单程(Single Route) 2h30min → 南湖码头(Nanhu Wharf)
半湖(Short Route)：镜泊山庄码头(Jingpo Villa Wharf) 往返(Round Route) 1h45min → 北湖码头(Beihu Wharf)

火山口森林路线 Forest in Crater route

镜泊湖北门(North Gate of Jingpo Lake) (27km/30min) → 火山口森林正门(Main Entrance of Forest in Crater)
火山口森林正门(Main Entrance of Forest in Crater) (23km/25min) → 火山口森林广场(Forest in Crater Square)

导览图 Guide Map 镜泊湖世界地质公园管委会提供 Provided by Jingpohu UGGp Management Committee

峡谷瀑布 Waterfalls in Jingpo Valley 镜泊湖世界地质公园管委会提供 Provided by Jingpohu UGGp Management Committee

溪韵 The Charm of Streams
镜泊湖世界地质公园管委会提供 Provided by Jingpohu UGGp Management Committee

瀑布飞人 Flying Man in Jingpohu 摄影：朱益民 Photographed by Zhu Yimin

峡谷秋色 Autumn Scenery of Jingpo Valley 镜泊湖世界地质公园管委会提供 Provided by Jingpohu UGGp Management Committee

紫菱湖湿地（1）Ziling Lake Wetland （1）镜泊湖世界地质公园管委会提供 Provided by Jingpohu UGGp Management Committee

紫菱湖湿地（2）Ziling Lake Wetland （2）镜泊湖世界地质公园管委会提供 Provided by Jingpohu UGGp Management Committee

冰雪世界（1）Ice and Snow World（1）镜泊湖世界地质公园管委会提供 Provided by Jingpohu UGGp Management Committee

冰雪世界（2）Ice and Snow World（2）
镜泊湖世界地质公园管委会提供 Provided by Jingpohu UGGp Management Committee

镜泊湖之秋（1）Autumn Scenery of Jingpo Lake（1）
镜泊湖世界地质公园管委会提供 Provided by Jingpohu UGGp Management Committee

镜泊湖之秋（2）Autumn Scenery of Jingpo Lake（2）
镜泊湖世界地质公园管委会提供 Provided by Jingpohu UGGp Management Committee

镜泊湖之秋（3）Autumn Scenery of Jingpo Lake（3）镜泊湖世界地质公园管委会提供 Provided by Jingpohu UGGp Management Committee

02 神奇镜泊

MIRACULOUS JINGPOHU

镜泊湖，在《汉书·地理志》中被称为湄沱河，唐高宗时称阿卜湖，唐玄宗时称忽汗海，辽时称扑莹水，明志始称镜泊湖。

镜泊湖世界地质公园位于张广才岭和老爷岭两山脉之间，地势西北高东南低。公园西北部分布着大大小小的16个火山口，山势起伏较大，东南部地势较为平缓。区内地貌主要是中山陡坡、中低山和低山丘陵平原。

镜泊湖地区属温带大陆性湿润季风气候，四季分明。春季干燥多风；夏季温热多雨；秋季短促，日照充足；冬季寒冷漫长。

宁静秀美的湖泊，气势磅礴的弧形瀑布，繁盛茂密的地下森林，巨大的火山塌陷坑，幽深的熔岩隧道等自然奇观，共同组成了神秘多姿的镜泊世界。

Jingpo Lake was named as Meituo River in *Geographica of History of Han Dynasty*, Abu Lake under the reign of Emperor Gaozong of Tang Dynasty, Huhanhai Lake under the reign of Emperor Xuanzong of Tang Dynasty, Puyingshui Lake in Liao Dynasty, and called as Jingpo Lake since Ming Dynasty.

Jingpohu UGGp is located between Zhangguangcailing and Laoyeling, with terrain low in the southeast and high in the northwest. In the northwest of the park, there are 16 craters of different sizes, thus the mountains there are undulating while the terrain in the southeast is flat. The topography in the Geopark is characterized by medium-mountains with steep slopes, middle-low mountains, low hills and plains.

Jingpohu area enjoys a humid continental monsoon climate featuring four distinct seasons. It is dry and windy in spring, warm and rainy in summer. Autumn here is short, but sunlight is sufficient, while the winter is long and cold.

The wonderland of Jingpohu has many natural wonders, such as highlighting serene and elegant lakes, grand and magnificent arc waterfalls, and dense underground forest, huge volcano collapse pits, deep lava tubes and so on.

飞瀑彩虹 Waterfalls and Rainbow
摄影：靳国军 Photographed by Jin Guojun

湖水的“画” The “Painting” of Lake Water 摄影：刘洪群 Photographed by Liu Hongqun

变换 The Transformations 摄影：吴继学 Photographed by Wu Jixue

湖心岛晨 The Morning of Mid-lake Island
摄影：朱益民 Photographed by Zhu Yimin

峡谷雾凇 The Rime in the Valley 镜泊湖世界地质公园管委会提供 Provided by Jingpohu UGGp Management Committee

秋枫（1）Maple Leaves in Autumn（1）
镜泊湖世界地质公园管委会提供
Provided by Jingpohu UGGp Management Committee

秋枫（2）Maple Leaves in Autumn（2）
镜泊湖世界地质公园管委会提供
Provided by Jingpohu UGGp Management Committee

秋色 Autumn 摄影：孙江林 Photographed by Sun Jianglin

火山渣 Cinder/Scoria 摄影：李江风 Photographed by Li Jiangfeng

火山弹 Volcanic Bomb 镜泊湖世界地质公园管委会提供 Provided by Jingpohu UGGp Management Committee

冰洞 Ice Cave 镜泊湖世界地质公园管委会提供 Provided by Jingpohu UGGp Management Committee

03 地质记忆

GEOLOGICAL MEMORY

镜泊湖世界地质公园内拥有丰富多样的地貌类型，有神秘奇特的火山地貌，有造型各异的花岗岩地貌，还有壮阔雄奇的构造地貌。这些地貌见证着镜泊湖地区的地质历史和地质变迁。

Jingpohu UGGp has a rich variety of landforms, including peculiar volcanic landform, granite landform of different shapes, as well as magnificent tectonic landform, all of which have been witnessing the geological history and changes of Jingpohu region.

熔岩台地 Lava Plateau 镜泊湖世界地质公园管委会提供 Provided by Jingpohu UGGp Management Committee

距今约8亿—5.4亿年的新元古代晚期，区域上发生了一次大规模造山运动，形成了张广才岭隆起及张广才岭期侵入岩，使镜泊湖所在的区域形成变质岩和岩浆岩为主的古老基底。镜泊湖区的小孤山出露的花岗闪长岩，就是这个时期侵入的岩石。

距今约4.1亿年的早泥盆世晚期，因海水侵漫，在小北湖一带形成海–陆相碳酸盐岩–碎屑岩组合。距今约3.7亿年的晚泥盆世，地壳张裂形成老秃顶子组酸性火山岩，晚泥盆世末期（距今约3.6亿年），地壳上升，沉积间断。

距今约2.3亿年的中生代三叠纪晚期，地壳张裂形成罗圈站组中酸性火山岩，火山活动间歇期沉积了轻变质的泥质板岩和砂质板岩。印支期（距今约2亿年）岩浆活动波及区内，产出岩基、岩株、岩脉等。

白垩纪晚期（距今约1亿—0.6亿年），敦密断裂带形成了张广才岭–老爷岭山地和断坳陷山间盆地的构造地貌形态，盆地中沉积了海浪组河湖相陆源碎屑岩地层。

距今约6500万年新生代古近纪，喜马拉雅山运动使山地增高，盆地相对下降。新近纪中期（距今2300万—1500万年）沉积了上门子组河湖相砂岩、砂砾岩间夹湖沼相的粉细砂岩、泥岩和煤层。新近纪晚期（距今1500万年），火山活动在老爷岭山区表现为船底山组玄武岩喷溢，玄武岩厚100余米。

In the late Neoproterozoic about 800 to 540 million years ago, a large-scale intensive orogeny occurred, thus forming Zhangguangcailing uplift and Zhangguangcailing intrusive rock which led to the formation of the ancient base mainly consist of metamorphic rock and magmatic rock in Jingpohu region. For example, the granodiorite exposed in Xiaogu Mountain in Jingpohu region is the intrusive rock formed in that period.

In the late Early Devonian about 410 million years ago, a marine-continental facies carbonatite and clasolite assemblage was formed in Xiaobei Lake due to the invasion of sea water. About 370 million years ago in the Late Devonian, crust rifted to formed acidic volcanic rock of Laotudingzi group. At the end of the Late Devonian (about 360 million years ago), the deposition was interrupted due to crustal uplift.

About 230 million years ago in the Late Triassic Mesozoic, crustal rifted to formed medium acidic volcanic rocks of Luoquanzhan group. During the volcanic pause, there were low-grade metamorphic argillite and silty slates deposited. During the Indosinian (about 200 million years ago) the area was affected by magmatic activities, thus forming batholith, rock stock, dike, etc.

In Late Cretaceous (about 100 to 60 million years ago), the tectonic geomorphology of the Zhangguangcailing - Laoyeling mountains and depression intramontane basin was formed by the Dunmi faulted zone, in the meantime, the fluvial and lacustrine facies of the Hailang group and the continental clastic strata have been deposited in the basin.

In the Cenozoic Paleogene (about 65 million years ago), the Himalayan movement uplifted the mountains and caused the basins to subside relatively. In the middle Neogene (23 million to 15 million years ago), the fluvial and lacustrine facies sandstone of Shangmenzi group, glutenite embedded in or between fine silt, mudstone and coalbed of lacustrine-bog facies were deposited. In the late Neogene (15 million years ago), the volcanic activity is characterized by the basalt spillage of Chuandishan group in Laoyeling mountain area. The thickness of basalt is more than 100 meters.

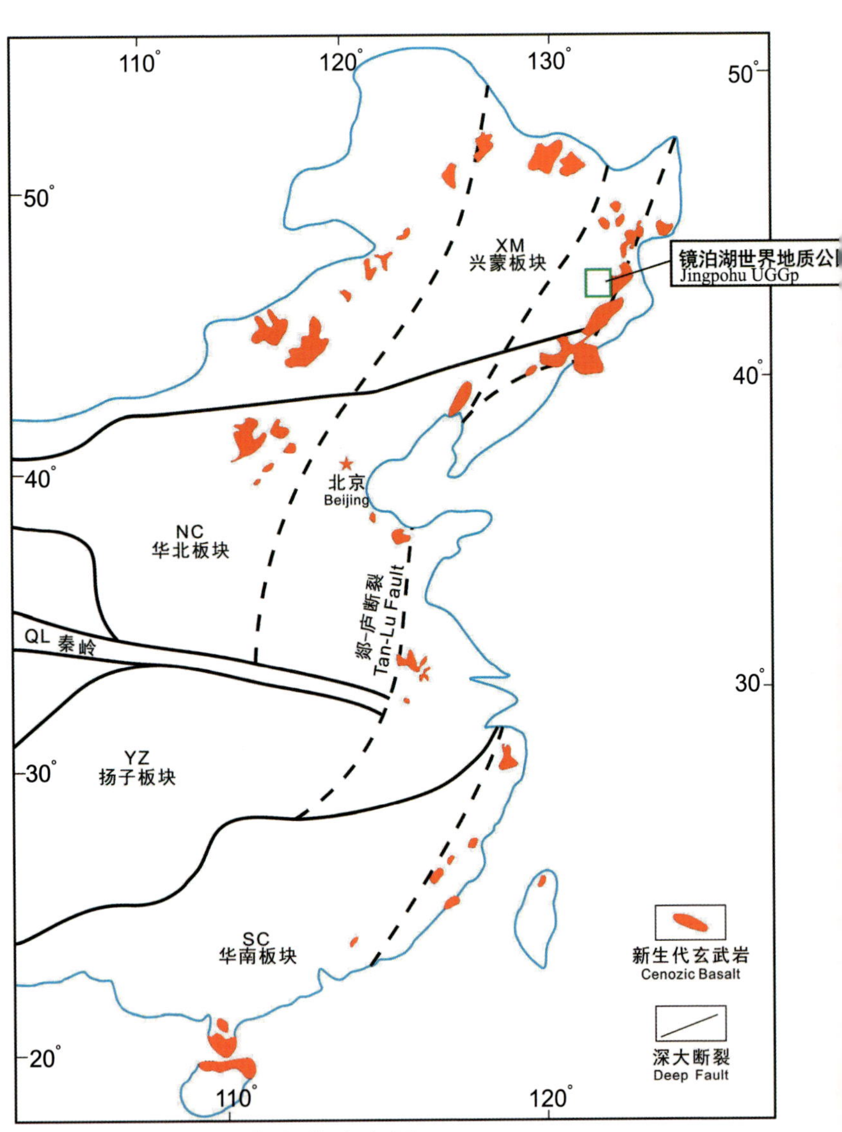

中国东部新生代岩浆岩带分布图
Distribution Map of Cenozoic Magmatic Belt in Eastern China
镜泊湖世界地质公园管委会提供
Provided by Jingpohu UGGp Management Committee

第四纪（距今258万年）以来，在山前盆地和山间沟谷中形成了河漫滩、阶地、台地及倾斜平原等地貌形态。同时火山活动频繁，相继发生了镜泊早期、中期、晚期和近期玄武岩喷发，并以裂隙-中心式喷发为主，形成16处火山口。

Since the Quaternary (2.58 million years ago), geomorphologic landforms such as floodplain, terrace, plateau and sloping plain have been formed in the foreland basin and intermontane gully. At the same time, the volcanic activity was frequent, which led to the consecutive occurrence of the early, middle, late and recent basalt eruptions in Jingpohu region. With 16 craters formed, the volcanic activity was mainly in the form of fissured and central eruption.

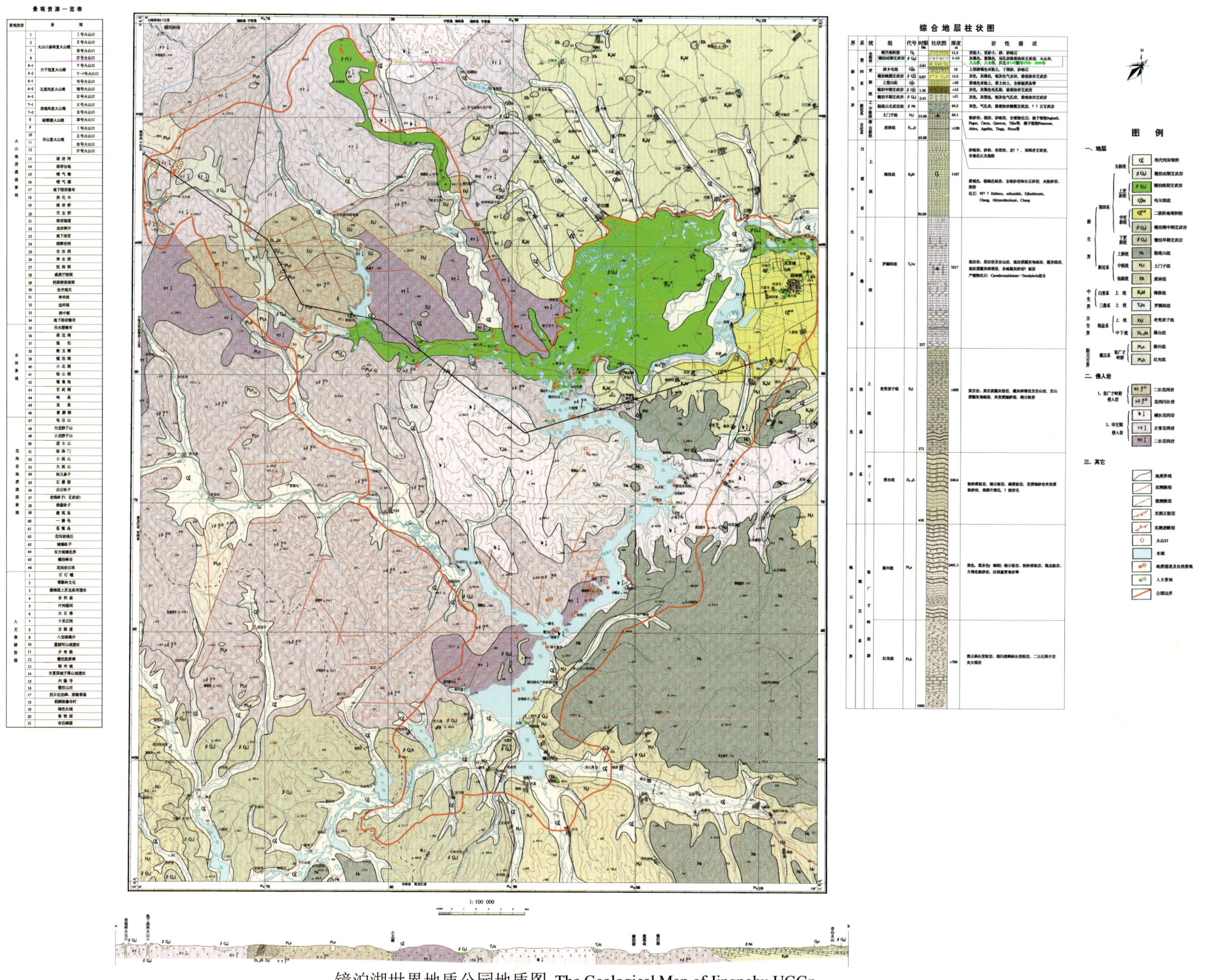

镜泊湖世界地质公园地质图 The Geological Map of Jingpohu UGGp

镜泊湖世界地质公园管委会提供 Provided by Jingpohu UGGp Management Committee

火山地貌
Volcanic Landform

镜泊湖地区以火山地貌为主，火山口森林地区和蛤蟆塘地区为全新世玄武岩浆喷发，其中火山口森林地区为碧玄岩、碱性玄武岩和粗面玄武岩，而蛤蟆塘地区为碱玄岩，并含有丰富的歪长石、金云母巨晶和世界上少见的钛角闪石巨晶。

Volcanic landform is dominant in Jingpohu region. Crater forest area and Hamotang area are formed by volcanic eruption in Holocene, in which the crater forest area is composed of basanite, alkali-basalt and trachybasalt. While the Hamotang area is composed of alkali-basalt, containing abundant anorthoclasite, phlogopite megacrysts and kaersutite megacrysts which are rare in the world.

消失的熔岩河 The Disappeared Lava River
镜泊湖世界地质公园管委会提供
Provided by Jingpohu UGGp Management Committee

构造地貌
Tectonic Landform

公园在区域地质构造上，处于西伯利亚板块与中朝板块之间，巴尔喀什-内蒙古-佳木斯联合板块中的布列亚-佳木斯微板块的西南缘，临近辽冀蒙板块与松辽微板块结合部位。强烈的构造活动，造就了区内褶皱山脉、断陷盆地、断裂三角面等构造地貌。

The geopark is located between the Siberian plate and the Sino-Korean plate, on the southwestern margin of Breya-Jiamusi micro-plate of the Balkash-Inner Mongolia-Jiamusi joint plate. It is in the vicinity of the junction zone of Liao-Ji-Meng plate and Song-Liao micro-plate in terms of the regional geological structure. The strong tectonic activity has resulted in structural landforms including fold mountains, fault basin, fault facets and so on.

构造地貌 Tectonic Landform
镜泊湖世界地质公园管委会提供
Provided by Jingpohu UGGp Management Committee

白石砬子 White Rock Cliff 镜泊湖世界地质公园管委会提供 Provided by Jingpohu UGGp Management Committee

花岗岩地貌
Granite Landform

公园内花岗岩出露面积1143km^2，主要岩性为6亿多年前张广才岭期的二长花岗岩、花岗闪长岩和2亿多年前印支期正长花岗岩、碱长花岗岩和二长花岗岩。在亿万年的地质历史时期里，区内的花岗岩经历了构造、侵蚀、风化等内、外动力地质作用，而形成了今天各具特色的珍贵的花岗岩地质遗迹景观。

The outcropped granite covers an area of 1143 km^2 in the geopark, with the main lithology of granite being monzonite granite and granodiorite in Zhangguangcailing stage of over 600 million years ago as well as syengranite, alkalic feldspar granite and monzonite granite in the Indosinian of over 200 million years ago. In the geological history of billions of years, the granite in the geopark has undergone the internal and external dynamic geological process such as tectonic action, erosion and weathering, thus forming today's distinctive and precious granite landforms.

蘑菇石 Mushroom Stone
镜泊湖世界地质公园管委会提供
Provided by Jingpohu UGGp Management Committee

石臼坑 Stone Mortar
镜泊湖世界地质公园管委会提供
Provided by Jingpohu UGGp Management Committee

花岗岩 Granite
镜泊湖世界地质公园管委会提供
Provided by Jingpohu UGGp Management Committee

珍珠门 Pearl Gate 镜泊湖世界地质公园管委会提供 Provided by Jingpohu UGGp Management Committee

04 火山奥秘

MYSTERY OF VOLCANOES

镜泊火山群地处欧亚大陆板块东部、老爷岭地块区张广才岭-太平岭边缘隆起带的南部。新生代以来，受滨太平洋板块向西俯冲的应力效应影响，火山活动频繁，岩浆沿断裂构造的交汇裂隙处喷溢，形成了16个大小不一的火山口。

Jingpo volcanic group is located in the eastern Eurasian plate and the southern part of the Zhangguangcailing-Taipingling uplift zone at Laoyeling block. Since the Cenozoic, this area has been influenced by the westward subduction stress of Pacific plate, and has gone through the process of transformation from extensional tectonic environment to compression environment. The volcanic activities have been frequent here, and the lava spilled along the fault structure fissure at the junction zone, thus forming sixteen craters of different sizes in the geopark.

I号火山口 Crater No. I 摄影：任世君 Photographed by Ren Shijun

火山口森林复火山 Forest in Crater Composite Volcanic Landform 摄影：李祥瑞 Photographed by Li Xiangrui

据地质学研究，镜泊火山群在距今1.2万年、8300年和5140年有过三次喷溢活动。其喷出的熔浆奔泻于山谷之间，形成了累计长度大于27千米的熔岩隧道，堵塞了牡丹江古江道，形成了大型火山熔岩堰塞湖——镜泊湖，以及众多的小型堰塞湖。留下了典型、稀有、系统、完整的火山地质遗迹景观——火山口、熔岩流、喷气碟、熔岩台地、熔岩隧道、火山碎屑物、火山岩等，因而成为世界上不可多得的天然火山岩博物馆。

According to geological research, Jingpo volcanic group erupted three times at 12 000, 8300 and 5140 years ago respectively. The erupted lava ran through the valleys and formed lava tubes with a cumulative length of more than 27 km; Volcanic eruptions obstructed the ancient watercourse of Mudanjiang River, and formed large volcanic lava dammed lake——Jingpo Lake together with numerous small lava dammed lakes, leaving typical, rare, systematic and complete landscape of volcanic geological heritage landscapes, including crater, lava flows, fumarolic cone, lava plateau, lava tunnel, pyroclastics, and volcanic rocks, thus Jingpohu UGGp becomes one of the rare museums of natural volcanic rocks in the worl

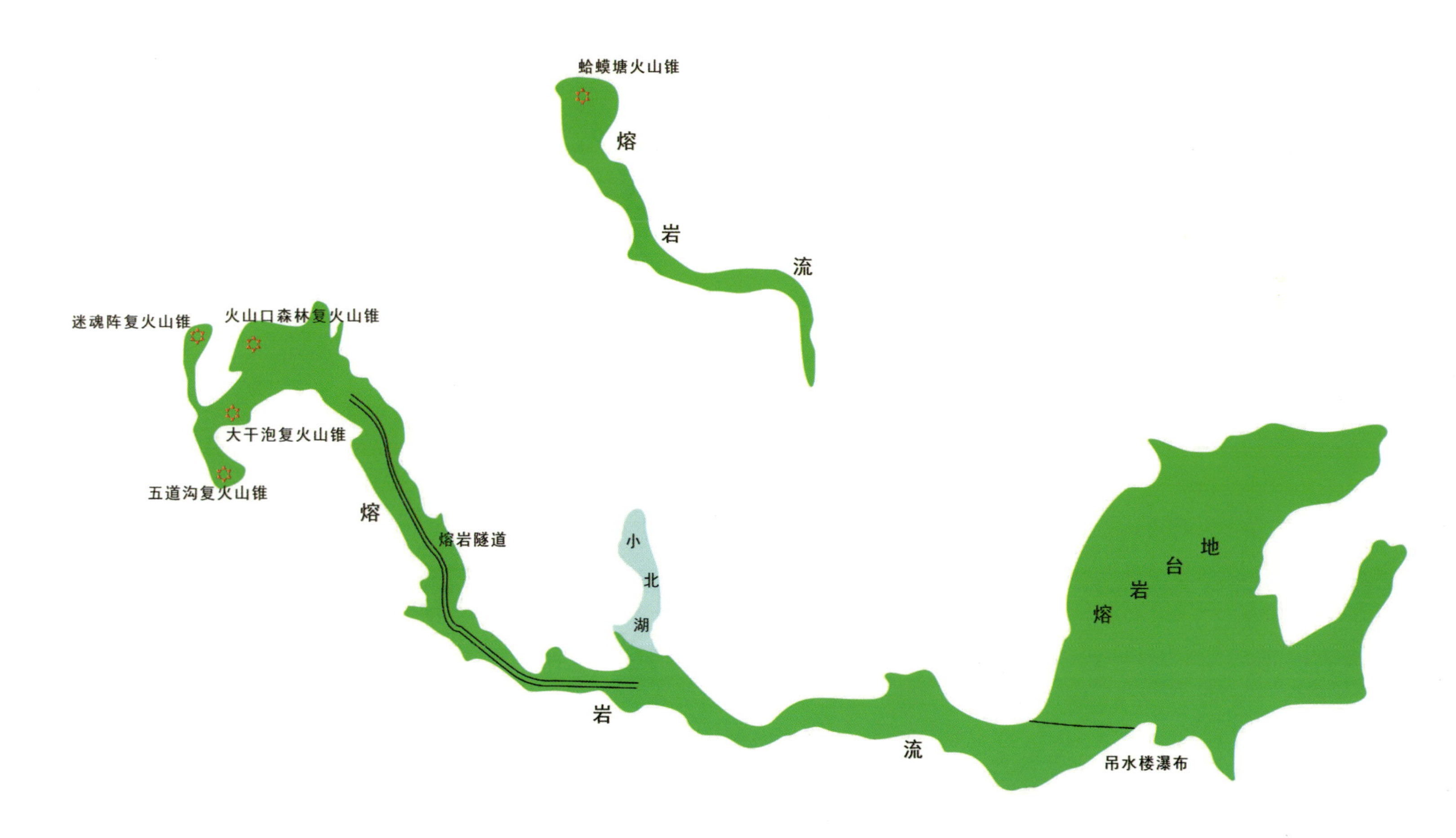

熔岩流分布示意图 Lava Flows Distribution Diagram
镜泊湖世界地质公园管委会提供 Provided by Jingpohu UGGp Management Committee

柱状节理（1）Columnar Joints（1）
摄影：王江 Photographed by Wang Jiang

柱状节理（2）Columnar Joints（2）
摄影：王江 Photographed by Wang Jiang

熔岩隧道（1）Lava Tunnel（1）
镜泊湖世界地质公园管委会提供
Provided by Jingpohu UGGp Management Committee

熔岩隧道（2）Lava Tunnel（2）
镜泊湖世界地质公园管委会提供
Provided by Jingpohu UGGp Management Committee

鼓丘 Drumlins 摄影：李江风 Photographed by Li Jiangfeng

镜泊石海 Jingpo Lava Stone Sea 摄影：李江风 Photographed by Li Jiangfeng

波纹状熔岩 Wavy Lava 摄影：王江 Photographed by Wang Jiang

地下奇观 Underground Wonders
摄影：刘春 Photographed by Liu Chun

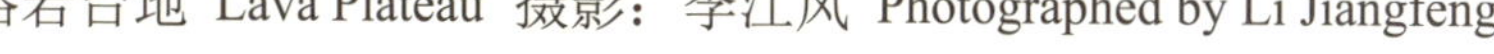

熔岩台地 Lava Plateau 摄影：李江风 Photographed by Li Jiangfeng

喷气碟 Fumarolic Cone 摄影：李江风 Photographed by Li Jiangfeng

神羊春光——镜泊湖神羊洞 The Beautiful Spring Scenery of Shenyang Cave—Shenyang Cave in Jingpohu UGGp
摄影：王欣 Photographed by Wang Xin

千层岩 Multi-layered Rocks 摄影：李江风 Photographed by Li Jiangfeng

浮石 Pumice 镜泊湖世界地质公园管委会提供 Provided by Jingpohu UGGp Management Committee

熔岩桥 Lava Bridge 摄影：李江风 Photographed by Li Jiangfeng

绳状熔岩（1）Pahoehoe Lava（1）摄影：李江风 Photographed by Li Jiangfeng

绳状熔岩（2）Pahoehoe Lava（2）摄影：李江风 Photographed by Li Jiangfeng

熔岩流 Lava Flows 摄影：李江风 Photographed by Li Jiangfeng

熔岩乳 Lava Stalactites 摄影：李江风 Photographed by Li Jiangfeng

小北湖保护区熔岩湿地 Lava Wetland in Xiaobei Lake Reserve 摄影：李江风 Photographed by Li Jiangfeng

峡谷一绝 Scenery in the valley 摄影：宋玉坤 Photographed by Song Yukun

05 丽水美景

BEAUTIFUL WATERSCAPE

俗话说“水火不容”，但在镜泊湖这片神奇的土地上，由于火山喷发时熔岩流动堵塞牡丹江道，后熔岩气洞发生塌陷，牡丹江水不断涌入，形成了大大小小的湖泊和湿地，使镜泊湖水火相容并且孕育出独特的丽水美景。

地质公园内镜泊湖、小北湖、紫菱湖、鸳鸯池、吊水楼瀑布等宛如一颗颗明珠，镶嵌在这片熔岩台地之上，其中尤以镜泊湖、吊水楼瀑布和紫菱湖闻名暇尔。

蜿蜒绵延的镜泊湖在群山环抱中时而水平如镜时而微波荡漾，“褶曲湖山几复湾，云落清波若镜天”（陈雷：《镜泊湖上》），“湖光山色绿黛敷，峰回流转湖连湖”（钱俊瑞：《调寄添字浣溪沙·镜泊湖》），“人在镜中行，云影天光上下明”（鲁歌：《游镜泊湖调寄南乡子》）等都描绘了镜泊湖的无尽秀美之景。岛湾错落，峰峦叠翠，景色清秀，古迹隐约，尽揽春花、夏水、秋叶、冬雪于一湖。

As the saying goes , "water and fire are incompatible". However, it is the "fire and water" that forged the wonderland of Jingpohu UGGp in the geological history. The flow of lava blocked Mudanjiang riverway due to the volcanic eruptions. After the lava gas cave collapsed, water of Mudanjiang River continued to flow in, forming lakes and wetlands of different sizes and forming a unique and beautiful waterscape where "fire and water" are compatible in Jingpohu.

In the Geopark, Jingpo Lake, Xiaobei Lake, Ziling Lake, Mandarin Duck Pool and Diaoshuilou Waterfalls scatter in this lava plateau like shining pearls. The Geopark is especially famous for Jingpo Lake, Ziling Lake and Diaoshuilou Waterfalls.

Winding through the mountains, the Jingpo Lake is sometimes limpid reflecting the blue sky like mirror and sometimes ripples like a jade ribbon. With its beautiful scenery and perfect ecologic environment, Jingpo Lake has its own characteristics of the four seasons. Many famous poets have depicted the endless beauty of Jingpo Lake. The scenery in Jingpo Lake is elegant and beautiful featuring various attractions of islands and bays, green peaks with lush vegetation, historic sites hidden in the nature, with the spring flowers, summer waters, autumn leaves and winter snow adding a different flavor to its picturesque landscape.

晨雾半岛 Morning Fog Peninsula
摄影：朱益民 Photographed by Zhu Yimin

镜泊行舟 Boating at Jingpo Lake
摄影：刘灿岭 Photographed by Liu Canling

鸳鸯池 Mandarin Duck Pool 镜泊湖世界地质公园管委会提供 Provided by Jingpohu UGGp Management Committee

生态湿地 Ecological Wetland
摄影：李江风 Photographed by Li Jiangfeng

紫菱湖秋色（1）Autumn Scenery of Ziling Lake（1）
摄影：李江风 Photographed by Li Jiangfeng

紫菱湖秋色（2）Autumn Scenery of Ziling Lake（2）
摄影：李江风 Photographed by Li Jiangfeng

冬瀑晓雾 Winter Waterfalls and Morning Mist
摄影：朱益民 Photographed by Zhu Yimin

镜泊湖水 Jingpo Lake 镜泊湖世界地质公园管委会提供 Provided by Jingpohu UGGp Management Committee

湿地景观 Wetland Landscape 摄影：李江风 Photographed by Li Jiangfeng

镜泊胜景 The Fascinating Scenery of Jingpo Lake 摄影：朱益民 Photographed by Zhu Yimin

响泉 Echo Spring 摄影：李江风 Photographed by Li Jiangfeng

湿地植被 Wetland Vegetation 摄影：李江风 Photographed by Li Jiangfeng

镜泊湖的湿地（1）Wetland in Jingpohu（1）摄影：李江风 Photographed by Li Jiangfeng

镜泊湖的湿地（2）Wetland in Jingpohu（2）摄影：李江风 Photographed by Li Jiangfeng

镜泊湖的湿地（3）Wetland in Jingpohu（3）摄影：李江风 Photographed by Li Jiangfeng

花海 Flower Sea 摄影：李江风 Photographed by Li Jiangfeng

气吞山河 Majestic and Powerful 摄影：王志军 Photographed by Wang Zhijun

倾泻的金河 The Torrential Golden River 摄影：安丰生 Photographed by An Fengsheng

小北湖初秋 Early Autumn of Xiaobei Lake
摄影：朱益民 Photographed by Zhu Yimin

火山岩湿地 Volcanic Wetland 摄影：朱益民 Photographed by Zhu Yimin

紫菱湖之秋 Autumn in Ziling Lake 镜泊湖世界地质公园管委会提供 Provided by Jingpohu UGGp Management Committee

06 密林深处

DEEP IN THE FORESTS

独特的自然条件使镜泊湖世界地质公园形成了别具一格的生态景观，这里群山环绕、森林密布、水域辽阔、生物种类繁多，构成了一套完整、和谐的自然生态系统。

Benefited from the unique natural conditions, Jingpohu UGGp enjoys a unique ecological landscape. There are lofty mountains, dense forests, vast waters, and well-preserved biodiversity, which constitute a complete and harmonious natural ecosystem.

白桦林 White Birch Forest 摄影：孙立波 Photographed by Sun Libo

远东豹 *Panthera pardus orientalis* 摄影：李世荣 Photographed by Li Shirong

动物天堂
Animal Paradise

镜泊湖优越的自然生境为各种野生动物提供了栖息繁衍空间，园区内分布有国家一级重点保护动物7种，包括东北虎、远东豹、梅花鹿、东方白鹤、金雕、丹顶鹤和中华秋沙鸭。国家二级重点保护动物35种，包括马鹿、斑羚、水獭、花尾榛鸡、鸳鸯、黑鸢、白肩雕等。此外，辽阔的水域还孕育了鲫鱼、红尾、胖头、鳌花等50余种鱼类。其中最有名的当属红尾鱼，是镜泊湖水域特有的鱼种。红尾鱼肉质细嫩，味道鲜美，是上好的美味佳肴。

The excellent natural environment of the geopark is suitable for the habitation and breeding of various wild animals. There are 7 species of national first-level key protected animals in the geopark, including *Panthera tigris altaica, Panthera pardus orientalis, Cervus nippon, Grus leugeranus, Aquila chrysaetos, Grus japonensis and Mergus squamatus.* There are also 35 species of state-level key protected animals, including *Cervus elaphus, Naemorhedus caudatus, Lutra lutra, Bonasa bonasia, Aix galericulata, Milvus migrans* and *Aquila heliacal,* etc. In addition, the vast waters are also home to more than 50 species of fish including *Carassius auratus, Pseudaspius leptocephalus, Aristichthys nobilis, Siniperca chuatsi* and other species, among which the most famous is the *Pseudaspius leptocephalus*, a fish that is endemic in Jingpo Lake. With the tender meat and succulent taste after being cooked, *Pseudaspius leptocephalus* is indeed an excellent delicacy.

金雕 *Aquila chrysaetos* 摄影：张岭 Photographed by Zhang Ling

大雁的故事（1）The Story of the Wild Goose（1）摄影：任世君 Photographed by Ren Shijun

大雁的故事（2）The Story of the Wild Goose（2）摄影：任世君 Photographed by Ren Shijun

苍鹭 Goshawk 摄影：孙厚轶 Photographed by Sun Houyi

镜泊白鹭 The Egret in Jingpohu 摄影：孙红宇 Photographed by Sun Hongyu

秋色（1）Autumn Scenery（1）摄影：孙江林 Photographed by Sun Jianglin

大丽花 Dahlias 摄影：李江风 Photographed by Li Jiangfeng

格桑花 Gesang Flower 摄影：李江风 Photographed by Li Jiangfeng

植物乐园
Plants Paradise

园区内植物种类丰富，植被覆盖率达90%以上，森林覆盖率大于60%。园区典型植被是红松阔叶混交林和落叶阔叶林，主要有落叶松、红松、樟子松、紫椴、冷杉、色木槭、水曲柳等树种，尤其是分布着红松、黄粟（黄菠萝）、东北红豆杉、蒙古栎等珍贵树种。茂密的森林也带来了大量经济产品，如灵芝、人参、刺五加、五味子等中药和猴头菌、松口蘑（松茸）、松子、蕨菜、榛子等山珍。

With more than 60% of forest coverage and more than 90% of vegetation coverage, there grow rich species of plants in the geopark. The typical vegetation can be divided into *Pinus koraiensis* broad-leaved mixed forest and deciduous broad-leaved forest, which mainly include *Larix gmelinii, Pinus koraiensis, Pinus sylvestris, Tilia amurensis, Abies fabri, Acer mono Maxim, Fraxinus mandshurica*, and other species. What is noteworthy is the distribution of such precious species as *Pinus koraiensis, Phellodendron amurense Rupr, Taxus cuspidate*, and *Quercus mongolica*. The dense forests also cultivate countless forest by-products, such as *Ganoderma lucidum, Panax ginseng, Acanthopanax senticosus, Schisandra chinensis* and other Chinese medicinal plants as well as *Hericium erinaceus*, *Tricholoma*, *Pine nut*, *Pteridium aquilinum*, *Corylus heterophylla Fisch* and other delicacies.

秋色（2）Autumn Scenery（2）摄影：朱益民 Photographed by Zhu Yimin

夏日花海（1）Flower Sea in Summer（1）摄影：李江风 Photographed by Li Jiangfeng

夏日花海（2）Flower Sea in Summer（2）摄影：李江风 Photographed by Li Jiangfeng

地下森林 Underground Forest 摄影：李江风 Photographed by Li Jiangfeng

苔藓 Moss 摄影：李江风 Photographed by Li Jiangfeng

火山红叶林 Volcanic Red Leaf Forest 镜泊湖世界地质公园管委会提供 Provided by Jingpohu UGGp Management Committee

盛夏 Midsummer 镜泊湖世界地质公园管委会提供 Provided by Jingpohu UGGp Management Committee

07 人文传奇

LEGENDARY HUMANITY

镜泊湖地区历史源远流长，最早可追溯到原始社会。这里有莺歌岭、松乙桥村、后渔村、珍珠门、渤海镇等近10处原始社会村落遗址。盛唐时期由靺鞨族建立的地方政权渤海国，有着“海东盛国”的美誉，留下了上京龙泉府、城墙砬子、白花甸子、大朱家屯等建筑遗址，对研究1300年前的渤海国历史文化具有重要的参考价值。

镜泊湖秀丽的风光吸引了历代僧尼来此建寺，促进了宗教文化的发展，比较著名的有药师古刹、兴隆寺等，其中兴隆寺内的石灯幢是渤海国佛教石雕艺术中的代表性杰作。

镜泊湖优美的自然景观和湖光山色，吸引诸多文人墨客驻足留下传世的诗词佳作。我国老一辈无产阶级革命家刘少奇、朱德、董必武、叶剑英和邓小平等人也曾到此撰文赋诗。

镜泊湖地区聚集着满族、朝鲜族等少数民族，浓厚的萨满文化、独特的朝鲜民族风情是镜泊湖一道亮丽的风景线。红罗女、神仙洞等美丽的神话传说又给这片净土增添了一丝神秘，令人向往。

兴隆寺 石灯幢 Xinglong Temple Stone Lamp Standard
摄影：王志军 Photographed by Wang Zhijun

Jingpohu region enjoys a long history which can be traced back to the primitive society. There are almost ten village sites of primitive society including Yingge Hill, Songyiqiao Village , Houyu Village , Pearl Gate , Bohai Town and others. In the flourishing period of Tang Dynasty, Mohe people established Bohai kingdom, a local state that was entitled as the "Glorious Kingdom in the East of the Sea". Some architecture sites survived and remained including Relics of Captial Shangjing Longquanfu, Wall Cliff, Baihua Meadow and Dazhujiatun Relics, which offer important references for studying the history and culture of ancient Bohai Kingdom 1300 years ago.

The magnificent environment of Jingpo Lake has also attracted monks and nuns in past dynasties to build temples here, which also promoted the development of religious culture. Among the temples, Pharmacist Temple and Xinglong Temple enjoy great reputation, meanwhile, the Stone Lamp in Xinglong Temple is the masterpiece of Bohai Buddhist stone sculpture art.

Since ancient times, the beautiful natural landscape of Jingpohu has inspired poets and men of literature to compose their masterpieces. In addition, the proletarian revolutionists of older generations in China such as Liu Shaoqi ,Zhu De, Dong Biwu, Ye Jianying and Deng Xiaoping also wrote poems during their visit in Jingpohu.

Jingpohu region is home to Manchu, Korean and other ethnic minorities, highlighting rich shaman culture and unique Korean ethnic customs endow this region with a profound culture. In addition, the time-honored legends of Hongluonv, fairy cave and many others also add a sense of mystery to the breathtaking beauty of wonderland.

药师古刹冬景 Winter Scenery of Pharmacist Temple
镜泊湖世界地质公园管委会提供 Provided by Jingpohu UGGp Management Committee

1 历史遗迹 Historical Sites

渤海国上京龙泉府遗址 Relics of Captial Shangjing Longquanfu in Bohai Kingdom of Tang Dynasty 摄影：李江风 Photographed by Li Jiangfeng

红罗女文化园 Hongluonü Cultural Garden 摄影：李江风 Photographed by Li Jiangfeng

八宝琉璃井 Eight-Treasure Glaze-Tiled Pavilion 摄影：李江风 Photographed by Li Jiangfeng

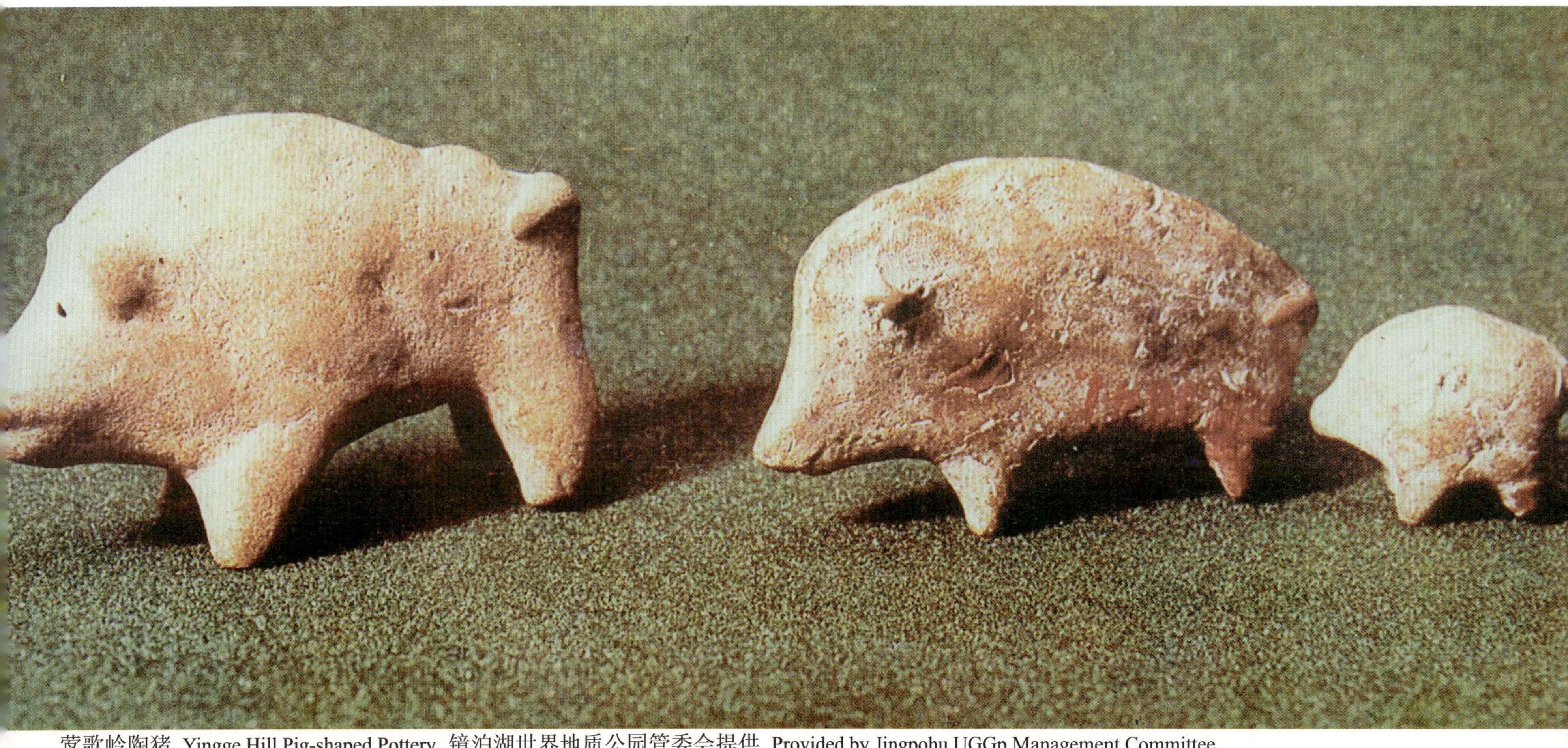

莺歌岭陶猪 Yingge Hill Pig-shaped Pottery 镜泊湖世界地质公园管委会提供 Provided by Jingpohu UGGp Management Committee

莺歌岭文物 Yingge Hill Antiques 镜泊湖世界地质公园管委会提供 Provided by Jingpohu UGGp Management Committee

药师古刹 Pharmacist Temple 摄影：李江风 Photographed by Li Jiangfeng

奇径碑园 Qijing Stele Garden
摄影：李江风 Photographed by Li Jiangfen

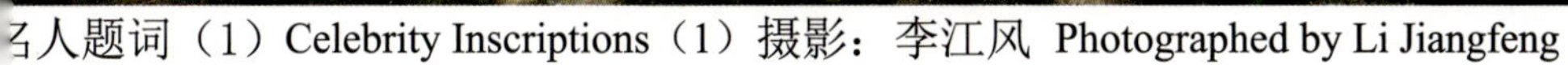

名人题词（1）Celebrity Inscriptions（1）摄影：李江风 Photographed by Li Jiangfeng

名人题词（2）Celebrity Inscriptions（2）摄影：李江风 Photographed by Li Jiangfeng

元首楼 Yuanshou Hotel
镜泊湖世界地质公园管委会提供 Provided by Jingpohu UGGp Management Committee

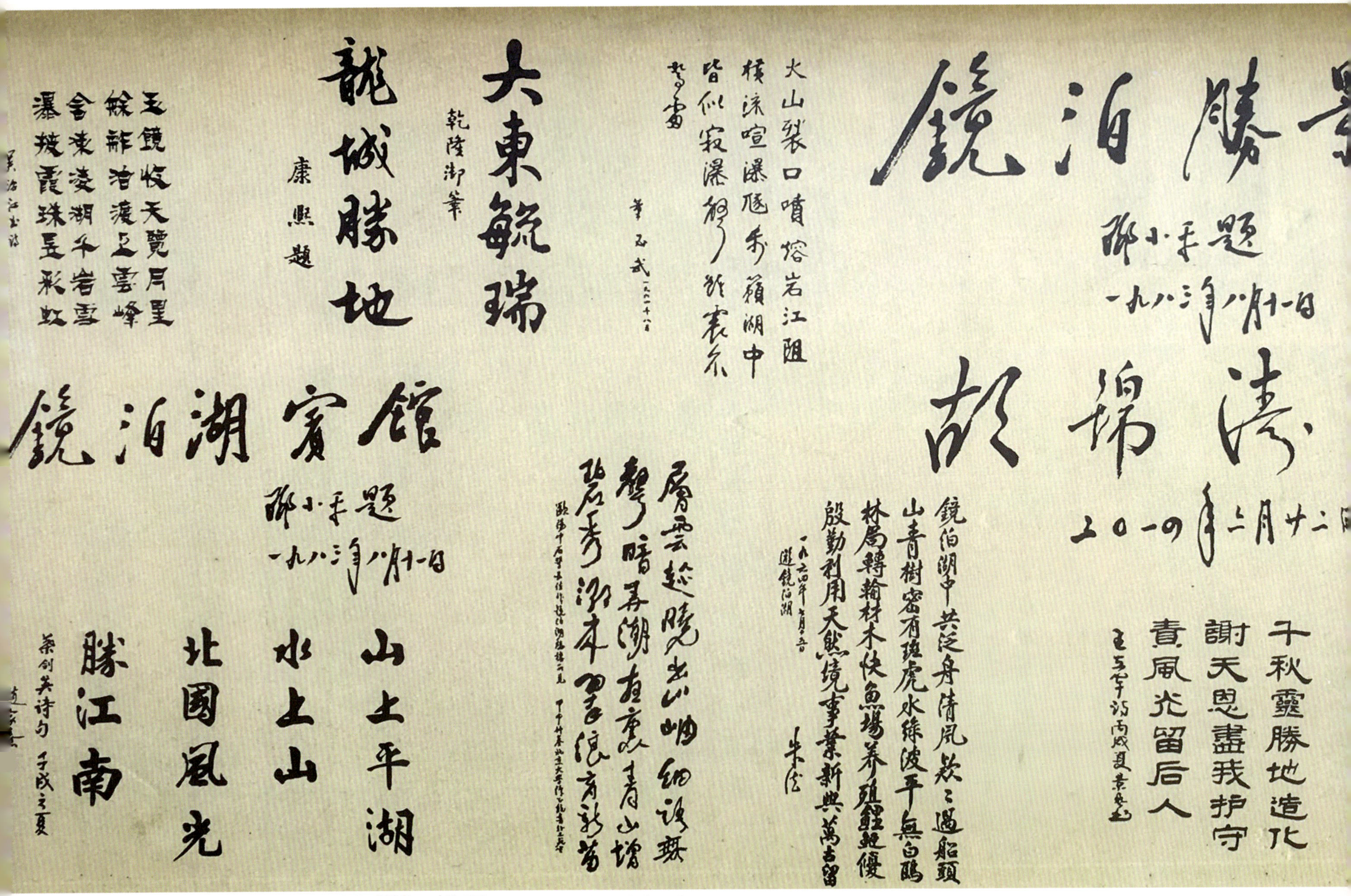

名人题词（3）Celebrity Inscriptions（3）摄影：李江风 Photographed by Li Jiangfeng

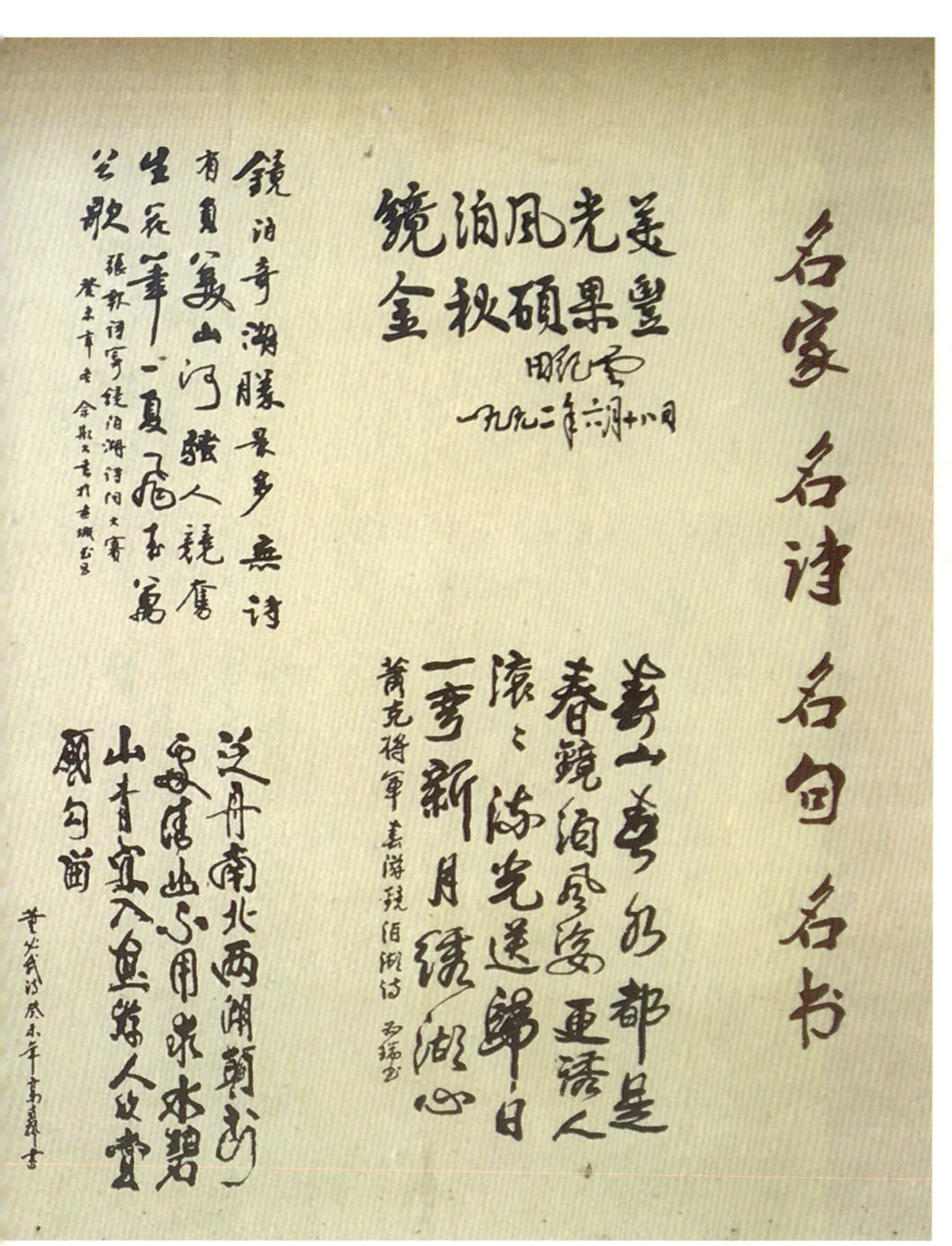

名人题词（4）Celebrity Inscriptions（4）
摄影：李江风 Photographed by Li Jiangfeng

宗教（1）Religion（1）镜泊湖世界地质公园管委会提供 Provided by Jingpohu UGGp Management Committee

宗教（2）Religion（2）镜泊湖世界地质公园管委会提供 Provided by Jingpohu UGGp Management Committee

农耕文化（1）Farming Culture（1）镜泊湖世界地质公园管委会提供 Provided by Jingpohu UGGp Management Committee

农耕文化（2）Farming Culture （2）镜泊湖世界地质公园管委会提供 Provided by Jingpohu UGGp Management Committee

响水大米 Xiangshui Rice
镜泊湖世界地质公园管委会提供 Provided by Jingpohu UGGp Management Committee

梅花鹿 *Cervus nippon*
镜泊湖世界地质公园管委会提供 Provided by Jingpohu UGGp Management Committee

牧马图 Horse Ranch Scene
镜泊湖世界地质公园管委会提供 Provided by Jingpohu UGGp Management Committee

渤海靺鞨绣 Bohai Mohe Embroidery 镜泊湖世界地质公园管委会提供 Provided by Jingpohu UGGp Management Committee

6 民俗文化 Folk Culture

戏曲 Chinese Opera 镜泊湖世界地质公园管委会提供 Provided by Jingpohu UGGp Management Committee

松木根雕摆件（1）Pine Root Carvings（1）镜泊湖世界地质公园管委会提供 Provided by Jingpohu UGGp Management Committee

俄罗斯套娃 Russian Matryoshka

松木根雕摆件（2）Pine Root Carvings（2）

山核桃装饰品 Pecan Ornaments

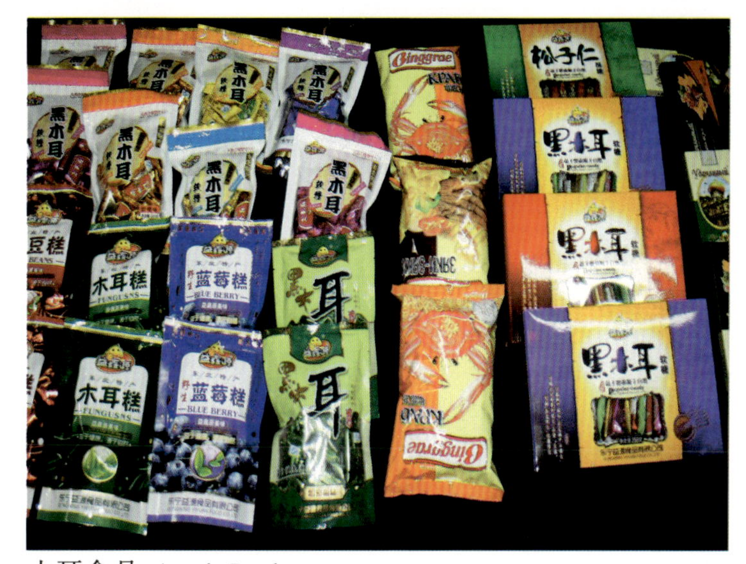

木耳食品 Agaric Products

鎏金首饰盒 Gilt Jewel Case

锡银首饰盒 Tin and Silver Jewel Case

08 持续发展

SUSTAINABLE DEVELOPMENT

自加入世界地质公园网络以来，镜泊湖世界地质公园在地质保育、地质科普教育、地质公园发展和促进区域社会经济进步等方面做出了卓越贡献，加强了对地质遗迹的保护和公园基础设施建设，完善了地质公园管理法规标准体系，开展了地学科普教育活动，举办了“冬捕节”“冬季阳光体育大会”“镜泊湖之夏旅游文化节”等节事活动，扩大了周边社区参与，推动了地方经济的可持续发展。镜泊湖世界地质公园积极参与联合国教科文组织世界地质公园网络的各项活动，与国内外地质公园开展交流与合作，分享了地质公园管理和地质遗迹保护的经验和方法。

镜泊湖是中国大地上一颗璀璨的明珠，是一部第四纪火山地质教科书，是上苍赐予人们的珍贵遗产。镜泊湖世界地质公园将用更加严格的管理、更加科学的方法、更加有力的措施保障公园的可持续发展，不断提升地质公园的影响力。让我们携起手来，共同保护这个美丽的家园，镜泊湖的明天将更加灿烂辉煌！

Since joining in the Global Geopark Network, Jingpohu UGGp has made remarkable contributions to the conservation of geohertiages, the education of geo-science, the development of geoparks and the promotion of regional social and economic progress. Meanwhile, it has strengthened the protection of geological relics and completed the establishment of the infrastructure in geoparks. It has also improved the standard system of geological park management laws and regulations, carried out activities of geosciences popular education. Furthermore, it has organized the “Winter Fishing Festival”, “Winter Sunshine Sports Games”, “Summer of Jingpohu Tourism and Culture Festival” and other festival activities, expanded the participation of the surrounding communities and promoted the sustainable development of local economy. Jingpohu UGGp actively participates in various activities of the UNESCO Global Geopark Network and carries out exchanges as well as cooperation with geoparks both at home and abroad to share experiences and methods of geological park management and protection on geological relics.

As a bright pearl in China, Jingpohu UGGp is not only a geological textbook of Quaternary volcanoes, but also a precious gift of the nature. Jingpohu UGGp will operate with more stringent management, more scientific methods and more effective measures to ensure the sustainable development of the geopark, while also constantly improving the influence of the geopark. Let us join hands to protect this beautiful geopark, and the future of Jingpohu will be more brilliant!

2014年镜泊湖世界地质公园中期评估 Revalidation of Jingpohu UGGp in 2014 摄影：李江风 Photographed by Li Jiangfeng

民族风情建筑 Folk Customs Architecture
摄影：李江风 Photographed by Li Jiangfeng

游船 Boat Sightseeing 摄影：朱益民 Photographed by Zhu Yimin

1 基础设施 Infrastructure

导览牌 Destination Boards 摄影：李江风 Photographed by Li Jiangfeng

洗手间 Restroom 摄影：李江风 Photographed by Li Jiangfeng

标识牌 Guide Signs 摄影：李江风 Photographed by Li Jiangfeng

自行车道 Cycling 摄影：李江风 Photographed by Li Jiangfeng

供租赁的自行车 Bicycle Rental 摄影：李江风 Photographed by Li Jiangfeng

科普长廊 Science Education Gallery 摄影：李江风 Photographed by Li Jiangfeng

游客服务中心 Tourist Service Center 摄影：李江风 Photographed by Li Jiangfeng

短道速滑比赛 Short Track Speed Skating Race 镜泊湖世界地质公园管委会提供 Provided by Jingpohu UGGp Management Committee

冬令营 Winter Camp 镜泊湖世界地质公园管委会提供 Provided by Jingpohu UGGp Management Committee

地学夏令营 Geo-camp 镜泊湖世界地质公园管委会提供 Provided by Jingpohu UGGp Management Committee

镜泊湖世界地质公园与牡丹江师范学院“科普教育基地”签约和揭牌仪式
Signing and Opening Ceremony of “Popular Science Education Base” Established by Jingpohu UGGp in Collaboration with Mudanjiang Normal University
镜泊湖世界地质公园管委会提供 Provided by Jingpohu UGGp Management Committee

地质主题社区（1）Geological Community（1）
镜泊湖世界地质公园管委会提供
Provided by Jingpohu UGGp Management Committee

地质主题社区（2）Geological Community（2）
镜泊湖世界地质公园管委会提供
Provided by Jingpohu UGGp Management Committee

地质主题社区（3）Geological Community（3）
镜泊湖世界地质公园管委会提供
Provided by Jingpohu UGGp Management Committee

社区发展（1）Community Development（1）摄影：李江风 Photographed by Li Jiangfeng

社区发展（2）Community Development（2）摄影：李江风 Photographed by Li Jiangfeng

山庄酒店 Hotel 摄影：李江风 Photographed by Li Jiangfeng

龙舟赛 Dragon Boat Race 镜泊湖世界地质公园管委会提供 Provided by Jingpohu UGGp Management Committee

捕鱼节（1）Fishing Festival（1）镜泊湖世界地质公园管委会提供 Provided by Jingpohu UGGp Management Committee

捕鱼节（2）Fishing Festival（2）镜泊湖世界地质公园管委会提供 Provided by Jingpohu UGGp Management Committee

图书在版编目（CIP）数据

水火“相容”——镜泊湖的地质故事 / 李江风等编；胡志红译. ——武汉：中国地质大学出版社，2018.7
ISBN　978-7-5625-4364-0

Ⅰ.①水…
Ⅱ.①李…　②胡…
Ⅲ.①地质—国家公园—介绍—宁安
Ⅳ.①S759.992

中国版本图书馆CIP数据核字（2018）第159958号

李江风　孙德志　马晓群　周学武　付崇华　万沙　陈梦婷　田野　编

水火“相容”——镜泊湖的地质故事　胡志红　译

责任编辑:舒立霞　版式设计:赵永景　责任校对:张咏梅

出版发行:中国地质大学出版社(武汉洪山区鲁磨路388号)　邮编:430074
电话:(027)67883511　传真:(027)67883580　E-mail:cbb@cug.edu.cn
经销：全国新华书店　http://cugp.cug.edu.cn

开本：889mm×1194mm　1/12　字数：414千字　印张：11.5
版次：2018年7月第1版　印次：2018年7月第1次印刷
印刷：武汉市籍缘印刷厂

ISBN978-7-5625-4364-0　定价：98.00元